# Creative Sparks for Busy Brains

# Creative Sparks for Busy Brains

## 110 ADHD-friendly prompts

## for creative flow with ChatGPT

By Viki Zangari

In collaboration with ChatGPT

Zangari Publishing

This book is a work of nonfiction. The content is provided for informational and educational purposes only. Readers are encouraged to evaluate and adapt ideas, strategies, and examples to their own circumstances. The author and publisher disclaim any liability for loss or damages resulting from the use of this material.

Portions of this book were developed through collaborative work with AI tools. All final content has been human-reviewed and shaped with care.

Published by

Zangari Publishing, Western Australia

ISBN: ISBN: 978-1-7643993-6-4

Cover design: By Viki Zangari (designed using Book Brush)
Interior design and formatting: Zangari Publishing (designed using Atticus)

Website: www.zangaripublishing.com

Email: zangaripublishing@gmail.com

# Contents

## Acknowledgements

This book came to life through a combination of curiosity, persistence, and the quiet belief that creativity deserves gentler pathways, especially for busy minds.

First, thank you to the readers. To those who live with ADHD, racing thoughts, interrupted focus, or creative energy that doesn't always move in straight lines, this book exists because of you. Your lived experience matters. Your imagination matters. And your way of thinking is worthy of tools that support it rather than suppress it.

I would also like to acknowledge the many researchers, clinicians, educators, and advocates whose work continues to shape our understanding of ADHD, executive function, creativity, and neurodiversity. While this book offers its own original framework, it is informed by the collective knowledge of professionals who have dedicated their careers to helping individuals better understand how their minds work.

A special acknowledgment goes to the evolving role of artificial intelligence as a creative support. Used thoughtfully, AI has the potential to lower barriers, reduce overwhelm, and make creative exploration more accessible. This book embraces that possibility with care, responsibility, and respect for the human voice at the center of every idea.

Finally, thank you to those who support creativity quietly, through encouragement, patience, and understanding. Creativity often grows not through pressure, but through permission.

This book is offered with that spirit

## Dedication

My motivation has always been to help those who struggle with ADHD. Over the years, I have made friends with many people who live with ADHD, and as awareness continues to grow, it is increasingly clear that ADHD touches far more lives than we once realized.

Today, it is widely understood that millions of people worldwide live with ADHD, and that the likelihood of knowing, loving, or working alongside someone with ADHD continues to rise each year.

This work is dedicated to all those navigating busy minds in a world that often asks for stillness. To those who think differently, create differently, and persist quietly.

You are not alone, and your way of being belongs.

# Introduction

If you're holding this book, chances are you have ideas, lots of them, but not always the time, energy, or mental space to catch them when they appear. Maybe your mind feels busy even when you want it to be quiet. Maybe inspiration arrives at inconvenient moments, then disappears just as quickly. Maybe you've been told to "focus," "sit still," or "finish what you start," without anyone explaining how that's meant to happen when your brain works the way yours does.

**This book was created for you.**

Creative Sparks for Busy Brains is not about forcing creativity or fixing the way you think. It's about working with your mind, not against it, especially if you live with ADHD, a fast-moving imagination, or a brain that resists rigid systems. You don't need long stretches of uninterrupted time. You don't need perfect focus. You don't need to feel motivated before you begin. What you need is a warm starting point.

## A Different Way to Approach Creativity

Creativity is often framed as something that requires discipline, silence, and control. But many busy minds thrive on curiosity, movement, and flexibility. When pressure is removed and expectations soften, ideas tend to surface more naturally. The prompts in this book are designed to: Reduce overwhelm. Spark curiosity instead of pressure. Offer structure without rigidity. Help ideas emerge in small, manageable moments. You are not meant to complete every prompt. You are not meant to move through the book in order. You are not meant to "do it properly." You are meant to **use** what helps and leave the rest.

## Why ChatGPT Is Part of This Book

For many busy brains, the hardest part of creating isn't the idea, it's getting started. ChatGPT is included here as a creative support tool, not a replacement for your voice or imagination.

**Used gently, it can:** Help you begin when your mind feels stuck. Offer suggestions when decision fatigue sets in. Act as a brainstorming partner that doesn't judge or rush. You remain the creator. The ideas remain yours. ChatGPT simply helps lower the barrier to entry.

### This Book Is Not a Test

There are no deadlines here. No expectation to be productive. Some days you might write a paragraph. Some days you might read a prompt and close the book. Creativity doesn't disappear when life gets busy, it just needs a different doorway. This book is one of those doorways. An Invitation

**Use this book when:** You want to create without committing to a big project. You need to think playfully again. You want support without pressure. Let this be a place you return to explore.

Your mind is busy, imaginative, and alive. This book exists to meet it there.

## Research-Informed Approach

"Creative Sparks for Busy Brains" is grounded in **widely accepted ADHD research and** strengths-based perspectives, translated into a creative, accessible format designed for everyday use. Rather than presenting clinical instruction, this book acts as a practical bridge, turning established ideas about ADHD into gentle creative supports that readers can adapt to their own lives. The approaches used in this book align with themes commonly discussed by leading ADHD clinicians and researchers, particularly in the areas of executive function, creativity, attention regulation, and cognitive support. Below are examples of how professional insights are interpreted into everyday language throughout this book.

## Executive Function & Structure

## (Informed by clinical and research perspectives on ADHD)

ADHD research frequently highlights challenges related to executive function, the mental processes involved in planning, starting tasks, sustaining focus, and organizing ideas.

**In simple terms:** This means that having a good idea is often not the problem, getting started or staying with it is.

**How this book applies that idea: Instead of asking readers to "sit down and be creative,"** this book uses: short prompts, Clear starting points. External structure. These supports reduce mental friction and make creativity feel more approachable.

**Strengths-Based ADHD Thinking (**Informed by strengths-focused clinical perspectives**)** Many professionals emphasize that ADHD is not just a challenge, it also comes with strengths such as divergent thinking, imagination, curiosity, and originality.

**In simple terms:** ADHD brains often make unexpected connections, especially when pressure is removed.

**How this book applies that idea:** Prompts are intentionally open-ended and playful, encouraging exploration rather than perfection. Readers are invited to follow curiosity, not outcomes.

**Cognitive Load & Mental Overwhelm** (Informed by psychological research) Research consistently shows that too many choices or unclear expectations can overwhelm working memory, particularly for ADHD minds.

**In simple terms:** Too many options can make it harder to choose any option.

**How this book applies that idea:** Each prompt provides: A clear focus. One starting direction Permission to stop early. This keeps creativity light, flexible, and manageable.

**External Supports & Creative Momentum** (Informed by executive-function research) ADHD research supports the use of external tools to reduce internal pressure and decision fatigue.

**In simple terms:** When your brain doesn't have to hold everything at once, creativity flows more easily.

**How this book applies that idea:** ChatGPT is used as a supportive creative partner, not a replacement for your voice, but a way to spark momentum, generate ideas, and lower the barrier to beginning.

**Encouragement, Not Correction;** "Creative Sparks for Busy Brains" does not attempt to diagnose, treat, or "fix" ADHD. Instead, it offers creative scaffolding, small supports that make imagination more accessible in the middle of real life. Readers who wish to explore the clinical and scientific foundations behind these ideas are encouraged to seek out books, lectures, and resources by respected ADHD professionals and researchers. Exploring multiple perspectives allows each reader to form an understanding that best suits their individual experience.

## About Research and Interpretation

The perspectives referenced in this book reflect **widely discussed themes** within ADHD research and clinical practice. They are presented here in simplified language to support creative understanding, not as clinical instruction or professional guidance. This book does not attempt to reproduce or teach any specific professional model, framework, or methodology. Instead, it offers a creative interpretation, translating general research concepts into prompts and practices that may feel more accessible in daily life. Readers are encouraged to explore original sources, professional publications, and qualified experts directly if they wish to deepen their understanding. Each individual's experience with ADHD is unique, and no single approach works for everyone.

## Recommended & Further Learning

This book has been created with deep respect for the lived experience of ADHD and is informed by contemporary research, clinical insight, and strengths-based perspectives shared by leading professionals in the field. While this work presents an original creative framework, readers who wish to explore ADHD further may find the following experts and resources valuable for continued learning.

## Recommended Experts

**Edward Hallowell, M.D.**
Psychiatrist, speaker, and author widely recognized for advancing a strengths-based understanding of ADHD. His work emphasizes creativity, connection, and reframing ADHD as a different way of thinking rather than a deficit.

**Ari Tuckman, Psy.D., MBA**
Clinical psychologist specializing in adult ADHD, executive function, and

practical strategies for everyday life. Known for translating research into accessible, actionable guidance for busy and creative minds.

**Russell Barkley, Ph.D.**
Clinical psychologist and researcher whose work on ADHD and executive function has shaped modern scientific understanding of attention, self-regulation, and cognitive support systems.

Readers interested in deepening their understanding of ADHD may wish to explore topics such as:

## Suggested Areas for Further Reading.

### Executive Function and Self-Regulation Primary, widely cited authors

**Russell Barkley, Ph.D.**; Foundational work on executive function as the core challenge in ADHD, particularly self-regulation, task initiation, and follow-through.

**Thomas E. Brown, Ph.D.**; Known for expanding executive function models to include activation, emotional regulation, and working memory. Strengths-Based ADHD Perspectives

### Primary authors and advocates

**Edward Hallowell, M.D.**; Prominent advocate for reframing ADHD as a difference that includes creativity, intuition, and high-energy thinking.

**John Ratey, M.D.**: Frequently collaborates with Hallowell; focuses on the brain's capacity for growth, movement, and creativity. Adult ADHD and Creative Cognition

## Key contributors

**Ari Tuckman, Psy.D., MBA;** Focuses on adult ADHD, motivation, identity, and real-world functioning, including creative and professional contexts.

**Stephen Faraone, Ph.D.;** Research leader on adult ADHD persistence and cognitive functioning across the lifespan. Practical Strategies for Focus and Follow-Through

## Applied and clinical authors

**Ari Tuckman, Psy.D:** Known for actionable, real-life strategies for task initiation, consistency, and follow-through.

**Russell Barkley, Ph.D.** Emphasizes external supports, accountability structures, and environmental design to compensate for executive function challenges. Idea Generation and Creative Support

## Relevant perspectives (indirect but aligned)

**Edward Hallowell, M.D.** Discusses creativity, imagination, and curiosity as natural ADHD strengths when pressure is reduced.

**Teresa Amabile, Ph.D.** Renowned creativity researcher whose work on intrinsic motivation and creative environments aligns with ADHD-friendly creative practices (not ADHD-specific, but highly applicable).

## Suggested Reading by Topic

## (For Independent Exploration)

The following book titles are offered as **optional resources** for readers who wish to explore ADHD research and related ideas more deeply. These works are widely cited and respected within clinical, educational, and research communities.

## Executive Function and Self-Regulation

**ADHD and the Nature of Self-Control,** by **Russell Barkley, Ph.D.**
A foundational work explaining executive function, self-regulation, and why starting, organizing, and sustaining tasks can be difficult for ADHD brains.

**A New Understanding of ADHD in Children and Adults,** by **Thomas E. Brown, Ph.D.**
Explores executive function challenges across the lifespan, including activation, emotional regulation, and working memory.

## Strengths-Based ADHD Perspectives

**Driven to Distraction, by Edward Hallowell, M.D. & John J. Ratey, M.D.**
A widely known work reframing ADHD as a difference that includes creativity, intuition, and high-energy thinking.

**Delivered from Distraction** by **Edward Hallowell, M.D. & John J. Ratey, M.D.**
Expands on strengths-based thinking while offering practical, compassionate guidance for adults with ADHD.

## Adult ADHD and Creative Cognition

**More Attention, Less Deficit** by **Ari Tuckman, Psy.D., MBA**
Focuses on adult ADHD, motivation, follow-through, and real-world functioning, including creative and professional contexts.

**Taking Charge of Adult ADHD** by **Russell Barkley, Ph.D.**
Discusses adult ADHD through an executive-function lens, emphasizing strategies that support daily functioning and momentum.

## Practical Strategies for Focus and Follow-Through

**The ADHD Advantage,** by **Dale Archer, M.D.**
Explores how ADHD traits can be leveraged productively with the right supports and structures.

**Finally Focused,** by **James Greenblatt, M.D.**
Offers practical, lifestyle-oriented strategies for improving focus and consistency (not creativity-specific but often referenced by adult readers).

## Idea Generation and Creative Support

**Creativity in Context,** by **Teresa Amabile, Ph.D.**
A foundational creativity text exploring how motivation, environment, and psychological safety support creative thinking.

**Originals,** by **Adam Grant, Ph.D.**
Examines how unconventional thinkers generate ideas and challenge norms — not ADHD-specific, but highly relevant to divergent thinking.

The books listed above are offered as optional resources for readers who wish to explore ADHD research, executive function, and creativity more deeply. They are not required to use or benefit from the prompts in this book.

## Publisher Disclaimer

This book is intended for educational and inspirational purposes only. It is not a substitute for professional medical, psychological, or therapeutic advice, diagnosis, or treatment.

The experts recommended in this publication are cited for educational context only. They have not reviewed, endorsed, or approved this book, nor is any formal affiliation implied.

All interpretations and applications presented here are the independent work of the author and publisher.

Readers are encouraged to seek qualified professional guidance for individual medical or mental health concerns. This book is best used as a creative resource alongside other forms of learning, reflection, and support.

## A Final Note to the Reader

There is no single "right" way to think, create, or imagine. This book offers one approach, gentle, flexible, and grounded in what we know about ADHD, while honoring that your experience is uniquely your own.

## How to Use This Book

"Creative Sparks for Busy Brains" is designed to be flexible, supportive, and pressure-free. There is no required order, no expectation to complete every prompt, and no "right" way to engage with the material. You are invited to use this book in the way that best supports your energy, attention, and curiosity. You May Use This Book By: Opening to any page that feels inviting. Choosing one prompt and stopping there. Skimming prompts to spark ideas without writing. Returning to the same prompt more than once. Using prompts alongside journaling, note-taking, or creative projects, Using ChatGPT as a creative partner to explore ideas at your own pace "Creative Sparks for Busy Brains" is not a course or a program. It is a creative companion, something to return to when inspiration feels blocked, overwhelming, or just out of reach.

## About the Research-Informed Content

Some sections of this book reference widely accepted ideas from ADHD research and creativity studies. These references are included to provide context and reassurance, not instruction. You are not expected to understand or study research in order to benefit from this book. If a concept resonates, you are welcome to explore it further through the suggested resources. If it does not, you are free to skip it entirely.

**What This Book Is and Is Not: This book is:** A creative support tool. A prompt-based idea starter. A flexible resource for busy, imaginative minds

**This book is not:** A diagnostic guide. A treatment plan. A replacement for professional support. Your creativity does not need fixing. This book simply offers gentle structure to help ideas surface more easily.

A Final Reminder: You are allowed to: Take breaks, Ignore prompts. Adapt language. Make this book your own. The most effective way to use this book is the way that feels kind, sustainable, and supportive to you.

# Chapter One

## The Doorway of Play & Imagination

### Where creativity begins with the smallest step

You don't stumble into this door it simply appears, as quietly as a thought you almost remembered. A soft glow rests at its edges, fluttering like candlelight in a cosy room. No expectations. No urgency. Just a gentle invitation. When you lean closer, you notice the air shifting, slightly warmer, slightly sweeter, as if the world behind the door already knows you're coming. This chapter is your welcome. It's the beginning of a journey guided by curiosity, not pressure. A place where ChatGPT helps you open the world one tiny detail at a time.

All you need to do is read the mini scene—paste the prompt—pick one detail from ChatGPT's response—use that detail as your next prompt

That's how your adventure unfolds, softly, simply, step by step. Now take a breath. Let the door open.

#### SPARK 1

#### The Door Opens Just a Little

**Mini scene.** The radiating door cracks open, sending a faint shimmer across the floor. A warm scent drifts out, something like spice, sunlight, or memory.

**Copy and Paste Prompt.** Describe the very first thing I see, smell, or hear when a radiating doorway opens into a low hum, imaginative world. Keep it comforting, simple, and sensory.

**Continue the journey—your next prompt.** Pick one detail from ChatGPT's answer and use it as your next prompt to guide the next step in your adventure.

## SPARK 2

### A Friendly Being Steps Forward

**Mini scene.** A small, kindly creature steps out from behind the doorframe. Its eyes are mellow, curious, like it's been waiting for you.

**Copy and Paste Prompt.** Create a friendly fantasy creature to greet me as I enter a slow-lit fun-loving world. Give it one charming quirk and a pleasant personality.

**Continue the journey—your next prompt.** Pick one detail from ChatGPT's answer and use it as your next prompt to guide the next step in your adventure.

## SPARK 3

### The World Slowly Reveals Itself

**Mini scene,** Beyond the threshold, the landscape forms piece by piece, like a painting coming alive in slow motion.

**Copy and Paste Prompt.** Describe a calm, imaginative landscape that forms slowly as I step through a glowing doorway. Include one unusual rule of this world.

**Continue the journey—your next prompt.** Choose the part of the landscape that draws your attention most, a colour, a movement, or a quiet rule of this world, and follow it as you step further into the scene.

## SPARK 4

### The Path Lights Up Beneath Your Feet

**Mini scene.** As you take your first step into the new world, the ground glows softly beneath your feet. It reacts to you, welcoming, playful, never overwhelming.

**Copy and Paste Prompt.** Describe a magical pathway that responds granularly when I step on it, lights, sounds, textures, or colours, in a steady and inviting way.

**Continue the journey—your next prompt.** Choose the part of the path's response that feels most welcoming, a light, a sound, or a texture, and let it guide your next step forward.

## SPARK 5

### Magic Follows Your Movements

**Mini scene.** A tiny magical effect trails behind you, maybe a sparkle, a warmth, a drifting light. It's subtle. inquisitively and playful.

**Copy and Paste Prompt.** Imagine a small magical effect that happens whenever I move through this world, something cute, subtle, or surprising.

**Continue the journey—your next prompt.** Choose the magical effect that feels most playful or intriguing and follow it as it trails you toward the next moment of your adventure.

## SPARK 6

### A Gentle Guide Appears

**Mini scene.** From behind a vibrant tree, a guide emerges. Not a teacher, just an ally who knows the paths here well.

**Copy and Paste Prompt.** Create a gentle guide character who joins me on this imaginative journey. Give them one useful skill and one amusing flaw.

**Continue the journey—your next prompt.** Choose the part of the guide that feels most reassuring or interesting like their skill, their flaw, or the way they move, and let it shape where you wander together next.

## SPARK 7

### A Message Finds You

**Mini scene.** A folded note appears in your hand. It hums quietly, as if written by the world itself.

**Copy and Paste Prompt.** Generate a short, friendly, slightly mysterious message that appears in my hand as I enter a magical realm. Keep it intriguing but comforting.

**Continue the journey—your next prompt.** Choose the part of the message that feels most comforting or intriguing and follow it as the world carefully responds around you.

## SPARK 8

### An Adventure-Version of You Appears

**Mini scene.** Ahead, a version of you steps into view, expressive, keen, and twinkling with possibility.

**Copy and Paste Prompt.** Create an adventure-version of me for this world. Describe their outfit, one magical ability, and one-off beat habit. Keep it friendly and imaginative.

**Continue the journey—your next prompt.** Choose the part of this adventure-version of you that feels most exciting or familiar, their outfit, ability, or quirky habit, and allow it to guide how you move forward in this world.

## SPARK 9

### Something From Home Blends into the World

**Mini scene.** A familiar item from your real life appears at your side, transformed slightly by the enchantment of this realm.

**Copy and Paste Prompt.** Bring one familiar object from my real life into this mystical world. Give it a harmless imaginative twist that fits the environment.

**Continue the journey—your next prompt.** Choose the part of the object's transformation that feels most meaningful or comforting and allow it guide how the familiar and the other world blend as you continue onward.

## SPARK 10

### The Sky Responds to Your Presence

**Mini scene.** Overhead, the sky shifts in light colours and patterns, as if acknowledging you.

**Copy and Paste Prompt.** Describe a sky that reacts delicately to my emotions, shifting colours, patterns, or shapes in a mildly way.

**Continue the journey—your next prompt.** Choose the change in the sky that feels most affirming or calm and let it linger as the world gently prepares you for whatever comes next.

## End of Chapter One

### The Doorway of Playful Beginnings

As you take a final look around the Atrium of Playful Beginnings, you realise how much has quietly awakened inside you. None of the sparks here were demanding. None asked you to be perfect. They simply invited you to notice, one colour, one shape, one moment of curiosity at a time. You've stepped through your first doorway, and in doing so, you've discovered something important: Your imagination responds beautifully when the world slows down and gives you just one detail to follow.

The shifting lights, the lively colours, the pleasant presence of your ChatGPT companion, they all worked together to show you that creativity doesn't start with big leaps. It starts with tiny sparks that feel possible. A soft breeze moves through the atrium, carrying a faint shimmer of light that gathers near your feet. It forms a pathway, subtle, glowing, inviting. It's not telling you what to do. It's simply leading you onward, toward the next miniature world waiting to unfold. You step onto the luminous path, carrying your first sparks with you, ready to see where this warm-hearted journey guides you next.

Chapter 2 will introduce a new place inside this world, and the first moment where ChatGPT becomes not just a path finder, but also a co-creator.

# Chapter Two

## The Hall of Echoed Ideas

### Where your thoughts become shapes, sounds, and gentle companions

As you walk deeper into the creative world, the light around you softens. The joyous radiance of the doorway fades into a cool, shimmering ambience, like moonlight reflecting off water.

In front, a vast hall begins to take shape. Not a grand hall. Not cold or intimidating. But spacious, open, curious, as if your ideas are meant to stretch their arms here.

The walls feel alive in a quiet way, peppered with soft glimmers that pulse like glowing stars. When you breathe, the lights respond, not brightly, not intensely, just enough to let you know this place listens.

This is the Hall of Echoed Ideas, a gentle chamber where thoughts don't disappear, they transform. Here, your imagination doesn't have to perform. It only has to whisper.

And ChatGPT becomes your companion in this space, helping you shape, grow, and explore whatever arrives.

Welcome to your second step on the journey.

## SPARK 1

### The Hall Responds to Your Presence

**Mini scene.** As you step inside, the radiant lights on the walls shift, forming iridescent trails that seem to follow your mood.

**Copy and Paste Prompt.** Describe how a fascinating hall reacts quietly to my presence as I enter, using colours, lights, or sounds that match the mode.

**Continue the journey—your next prompt.** Choose the way the hall responds that feels most in tune with you, a colour, a sound, or a quiet movement, and follow it as the space invites you further inside.

## SPARK 2

### A Whisper of an Idea Appears

**Mini scene.** A small glowing shape floats up from the floor. It's the size of a pebble, maybe a seed of a thought, waiting to grow.

**Copy & Paste Prompt.** Create a small glowing shape that represents a tiny idea forming in this hall. Describe its colour, movement, and mood.

**Continue the journey—your next prompt.** Choose the part of the glowing idea that feels most alive, its colour, its movement, or its mood, and let it guide how the thought begins to unfold.

## SPARK 3

### ChatGPT Joins You as a Soft Voice

**Mini scene.** You don't see a person, you simply sense a presence beside you, like a calm voice ready to help you explore this world.

**Copy and Paste Prompt.** Describe ChatGPT as a gentle, imaginative presence that helps me explore this hall. Give it a supple personality and one mischievous trait.

**Continue the journey—your next prompt.** Choose the part of this presence that feels most comforting or captivating, its tone, its frolicky trait, or the way it stays beside you, and permit it to guide how you explore the hall together.

## SPARK 4

### Your Idea Begins to Take Shape

**Mini scene.** The tiny luminescent idea pulses soothingly, with each pulse, it shifts, growing, stretching, humming quietly.

**Copy and Paste Prompt.** Show how the tiny glowing idea changes when I focus on it, shifting shape, colour, or texture in a gradual way.

**Continue the journey— your next prompt.** Choose the change in the idea that feels most interesting or promising, its shape, colour, or texture, and let it guide how the idea continues to form.

## SPARK 5

### A Second Idea Echoes from the Wall

**Mini scene.** One of the shimmering walls ripples, and an echo of your first idea appears, similar, but not the same.

**Copy and Paste Prompt.** Create an echo-version of my idea, similar to the first, but different enough to feel like a comical variation.

**Continue the journey—your next prompt.** Choose the variation that feels most playful or surprising and follow it as the echo invites the idea to evolve in a new direction.

## SPARK 6

### The Hall Reveals a Hidden Path

**Mini scene.** Between the glowing echoes, a narrow path appears, not straight, not winding, but tenderly inviting.

**Copy and Paste Prompt.** Describe a gentle hidden path that appears in the Hall of Echoed Ideas, shaped by my thoughts or emotions.

**Continue the journey—your next prompt.** Choose the part of the path that feels most inviting or emotionally resonant and have it guide where you wander next within the hall.

## SPARK 7

### A Thought You Forgot Returns

**Mini scene.** A small light drifts toward you from above. It feels like a half-forgotten idea, something little but meaningful.

**Copy and Paste Prompt.** Describe a sweetly, forgotten idea returning to me as a small glowing form. Keep it, light, sentimental, or quietly nostalgic.

**Continue the journey—your next prompt.** Pick one attribute from ChatGPT's answer and use it as your next prompt to guide your adventure.

## SPARK 8

### ChatGPT Helps You Expand the Scene

**Mini scene.** Your companion's presence radiates faintly, ready to help you shape the space around you.

**Copy and Paste Prompt.** Help me expand this scene. Describe one new detail in the hall, a sound, light, object, or motion, that deepens the atmosphere.

**Continue the journey—your next prompt.** Choose the new detail that deepens the atmosphere in a way that feels most natural or intriguing and allow it to delicately reshape how the hall unfolds around you.

## SPARK 9

### A Symbol Appears on the Floor

**Mini scene.** A pattern forms beneath your feet, delicate, radiant, shifting with your thoughts.

**Copy and Paste Prompt.** Describe a glowing symbol that appears beneath me in the hall, something abstract, soothing, and connected to imagination.

**Continue the journey—your next prompt.** Choose the part of the symbol that feels most calming or meaningful, and it guides how you stand, pause, or move within the hall.

## SPARK 10

### The Hall Opens to a New Realm

**Mini scene.** At the far end of the hall, a slow, wide opening forms, a new corner of this world calling gently for you.

**Copy and Paste Prompt.** Describe a new realm that reveals itself beyond the Hall of Echoed Ideas, soft, curious, and ready to explore.

**Continue the journey—your next prompt.** Choose the part of the new realm that feels most inviting or quietly exciting and allow it to linger as you prepare to step beyond the hall.

## End of Chapter 2

### The Soft Threshold of Story Seeds

As you take one final glance around the Story Seed Garden, you notice how the small details here have already begun to quietly shape your imagination. The seeds didn't rush you.  Every spark grew gently, at the pace your mind needed, one colour, one memory, one tiny story fragment at a time.

You explored miniature ideas, followed small sensory cues, and allowed yourself to wander lightly through the sprouts of imagination. Each moment reminded you that creativity can grow from the smallest beginning when you give yourself space to notice sparks instead of forcing outcomes. The air around the garden shimmers softly, as if acknowledging the seeds you've gathered and the ideas that have begun to bloom inside you.

Ahead, a faint glow marks a new doorway, not bright, not urgent, just quietly promising. It leads toward a realm where ideas swirl more freely, where imagination becomes playful, moving and shifting in gentle waves.

Your ChatGPT companion moves beside you, its presence steady and warmly curious, inviting you onward. You take one last look at the garden of beginning ideas and step toward the glow ahead, ready to meet the next realm where your creativity can expand in new, surprising ways.

# Chapter Three

## The Idea Current Atrium

### Where thoughts swirl, collide, and become possibility

As you step forward, the world begins to move differently. The air lifts. Colours loosen. The pleasant stillness of the Hall of Echoed Ideas gives way to motion, not chaotic, not overwhelming, but alive. Ideas drift like clouds, currents of thought passing through the space with perky energy. This is the Idea Current Atrium. Here, imagination behaves like weather. Thoughts gather, scatter, reform. Nothing stays fixed for long, and nothing needs to. This is a place for movement, variation, and creative momentum.

ChatGPT remains beside you, steady and curious, helping you notice patterns within the swirl rather than getting lost in it. You don't need to control the storm. You only need to follow one detail at a time.

#### SPARK 1

#### The First Current of Ideas

**Mini scene.** A pleasant breeze moves through the atrium, carrying faint shapes of colour and light, ideas just beginning to gather.

**Copy and Paste Prompt.** Describe the first current of an idea forming around me, using motion, colour, or light in an easy-going, imaginative way.

**Continue the journey—your next prompt,** Choose the part of the current that feels most alive or intriguing, a movement, a colour, or a flicker of light, and follow it as the ideas begins to take shape around you.

## SPARK 2

### Thoughts Begin to Spin

**Mini scene.** The air starts to rotate slowly, like ideas circling each other, inquisitive but unhurried.

**Copy and Paste Prompt.** Show how ideas begin to spin and move together in this atrium, delicately swirling without becoming overwhelming.

**Continue the journey—your next prompt.** Notice which part of the swirling motion draws your attention, the pace, the way ideas circle, or the calm within the movement, and let it carry the thought as it begins to form.

## SPARK 3

### A Colour Carries a Mood

**Mini scene.** One colour becomes brighter than the others, pulsing with a particular feeling.

**Copy and Paste Prompt.** Describe one colour in the idea current that carries a clear mood or emotion. Let it influence the space around me.

**Continue the journey—your next prompt.** Notice which colour draws you in, the emotion it carries or the way it changes the air and let its influence move through the storm as it continues to shift.

## SPARK 4

### A Sound Rises from the Current

**Mini scene.** A melodious sound emerges, not loud, not sharp, like a tone humming beneath the movement.

**Copy and Paste Prompt.** Describe a melodic sound that forms within the idea current and shifts as ideas move through the atrium.

**Continue the journey— your next prompt.** Notice how the sound shifts, its rhythm, its tone, or its quiet presence, and permit it to shape the way the current moves around you.

## SPARK 5

### An Idea Drifts Closer

**Mini scene.** One idea split's away from the current and floats nearer to you, enthusiastic and inviting.

**Copy and Paste Prompt.** Describe a single idea drifting closer to me. Give it a shape, texture, and movement.

**Continue the journey—your next prompt.** Notice what draws your curiosity, the subtle difference, the playful twist, or the way the idea feels familiar yet new, and let that closeness carry it in a fresh direction.

## SPARK 6

### The Current Responds to Attention

**Mini scene.** As you focus, the current subtly shifts, not intensifying, just reorganising itself around your awareness.

**Copy and Paste Prompt.** Show how the idea changes gently when I give attention to one detail, becoming mellower, happier, or more focused.

**Continue the journey—your next prompt.** Notice how the current responds to your focus, the calm that settles, the happiness that appears, or the clarity that forms, and permit that shift guide what unfolds next.

## SPARK 7

### A Pattern Emerges

**Mini scene.** Among the movement, a repeating rhythm begins to form, a pattern you hadn't noticed before.

**Copy and Paste Prompt.** Describe a gentle pattern that starts to emerge within the swirling ideas, something subtle but meaningful.

**Continue the journey—your next prompt.** Observe which part of the pattern begins to stand out, the rhythm, the repetition, or the feeling it carries, and let it delicately bring order to the movement around you.

## SPARK 8

### The Current Begins to Slow

**Mini scene.** The motion eases slightly. The colours soften. The energy begins to settle without disappearing.

**Copy and Paste Prompt.** Show how the idea current begins to slow and soften, transitioning toward calm while keeping a sense of creative energy.

**Continue the journey—your next prompt.** Sense which part of the slowing current feels most calming, the softened colours, the quieter motion, or the steady warmth that remains, and allow it to carry you into this gentler phase.

## SPARK 9

### A Clear Space Opens

**Mini scene.** At the centre of the atrium, a tranquil pocket of space forms, steady, open, breathable.

**Copy and Paste Prompt.** Describe a peaceful, clear space opening within the idea current, a place where thoughts feel lighter and more settled.

**Continue the journey – your next prompt.** Perceive which quality of this clear space feels most supportive, its openness, its steadiness, or the ease of breathing here, and rest with it as the storm settles fully around you.

### SPARK 10

### The Edge of the Current

**Mini scene.** Beyond the fading swirl, a warm, faint ray appears, the boundary between motion and rest.

**Copy and Paste Prompt.** Describe the edge of the idea atrium where movement fades into calm, warmth, reassuring, and ready to step through.

**Continue the journey—your next prompt.** Sense which part of this glowing edge feels most steady or reassuring and let it hold your attention as you pause at the boundary between movement and rest.

## End of Chapter 3

## The Edge of the Idea Current

As the last shapes of the Idea Current Atrium settle into a soft glow, you feel a gentle shift inside you, not loud, not dramatic, just a quiet awareness that your imagination is opening in new directions. Here in this space, you didn't have to chase inspiration. You only needed to follow a single detail and let it guide the next step of your adventure. The air begins to calm. The drifting colours fade into gentle pastel hues. And the current of ideas that swirled around you settles into a soft, steady glow.

Ahead, a warm haze opens into a quieter realm, a place where thoughts don't spin, where creativity breathes slower, where the world feels like a peaceful exhale. Your ChatGPT companion turns toward the quiet opening with a soft reassuring glow, inviting you to step into a calmer space where your mind can rest and reset. You take one last look at the sparkling shapes of the Idea Current, grateful for the energy it gave you, and then you follow the warm path forward, ready for the stillness and clarity waiting in the next chapter.

# Chapter Four

## The Sanctuary of Still Moments

### Where you pause, reset, and feel creativity return in small, steady waves

The realm beyond the Idea Current Atrium is quieter. The archway of glowing thought-shapes dissolves behind you, and you step into a place that feels like an exhale. A wide, friendly sanctuary opens around you, soft colours, warm air, faint echoes of distant wind chimes. Everything here seems designed not to impress you, but to comfort you. There is no need to rush. No need to create anything. No need to think quickly. Here, ideas come slower. softer and kindly. This is the Sanctuary of Still Moments, a place for grounding, resetting, and finding tiny sparks of motivation again, the kind that feels possible, not pressured. ChatGPT moves beside you quietly, matching your pace, ready to help you breathe through the world gently. Let's step deeper into calm.

#### SPARK 1

#### The Ground Supports You

**Mini scene.** You take one step, and the ground beneath you feels warm and steady, as if this realm is holding you up without effort.

**Copy and Paste Prompt.** Describe the ground in this sanctuary as something warm, steady, and supportive, include colours, textures, or sensations that feel grounding.

**Continue the journey—your next prompt.** Notice where the ground feels most comforting, its warmth, or the sense of being supported, allow that feeling to settle as you move deeper into the sanctuary.

## SPARK 2

### A Gentle Light Settles Around You

**Mini scene.** A soft glow wraps around you like a familiar blanket, not bright, not dim, just enough to feel safe.

**Copy and Paste Prompt.** Describe a gentle light that settles around me in this sanctuary, something calming, warm, and quietly reassuring.

**Continue the journey—your next prompt.** Notice which quality of the light feels most comforting, its warmth, its softness, or the way it quietly surrounds you, and allow that glow remain with you as the sanctuary holds you in calm.

## SPARK 3

### A Restful Bench Appears

**Mini scene.** A small bench forms from smooth stone and soft moss, as if the sanctuary itself wants you to pause.

**Copy and Paste Prompt.** Create a peaceful resting place in this sanctuary, a bench, cushion, stone, or shape. Make it feel inviting and safe.

**Continue the journey—your next prompt.** Notice which part of the resting place feels most welcoming, the texture beneath you, the way it supports your body, or the calm it offers, and allow yourself to linger there without needing to move on.

## SPARK 4

### A Slow Breeze Clears Your Mind

**Mini scene.** A breeze drifts through, cool, thoughtful, and slow. It doesn't bring ideas; it clears space for them.

**Copy and Paste Prompt.** Describe a soft breeze that clears mental clutter in a kindly way, focusing on sensations, sounds, or colours in the air.

**Continue the journey—your next prompt.** Sense which quality of the breeze feels most refreshing, its coolness, its movement, or the quiet it leaves behind, and let that clearing open a little more space inside you.

## SPARK 5

### A Small Source of Motivation Appears

**Mini scene.** A tiny glowing object appears beside you, not flashy but quietly energising.

**Copy and Paste Prompt.** Create a small, peaceful object that represents motivation returning, describe its glow, texture, and emotional feeling.

**Continue the journey—your next prompt.** Notice which quality of this small glow feels most encouraging, its warmth, its steadiness, or the quiet hope it carries, and allow that feeling to rest with you without asking anything in return.

## SPARK 6

### ChatGPT Offers a Tiny Encouraging Thought

**Mini scene.** Your companion's presence radiates for a moment, almost like a soft nod of reassurance.

**Copy and Paste Prompt.** Generate a tiny, gentle encouraging thought that fits this sanctuary, something soft, kind, and low-pressure.

**Continue the journey—your next prompt.** Notice which part of the encouragement settles most naturally, the words themselves, the feeling behind them, or the reassurance they bring, and let it stay with you quietly as you rest in this moment.

## SPARK 7

### The Environment Responds to Your Calm

**Mini scene.** As your breathing slows, the colours around you shift slightly, softening, deepening, glowing more warmly.

**Copy and Paste Prompt.** Describe how the sanctuary responds when I feel calm, small colour changes, soft movements, or warm sounds.

**Continue the journey—your next prompt.** Notice which change in the environment feels most soothing, the softened colours, the gentle movement, or the warm sound, and let that response quietly deepen your sense of calm.

## SPARK 8

### A Cup of Restorative Warmth Appears

**Mini scene.** A small cup materialises beside you, filled with something warm, fragrant, and calming.

**Copy and Paste Prompt.** Describe a warm drink that appears in this sanctuary, include scent, taste, temperature, and emotional effect.

**Continue the journey—your next prompt.** Notice which aspect of the warmth feels most comforting, the scent rising from the cup, the temperature in your hands, or the calm it brings, and let yourself savour that feeling without needing to do anything else.

## SPARK 9

### A Symbol of Renewal Forms Nearby

**Mini scene.** A sparkling symbol forms in the air, slow, luminous, pulsing tenderly.

**Copy and Paste Prompt.** Describe a lively symbol of renewal that appears in the sanctuary, its shape, meaning, and serene enchantment.

**Continue the journey—your next prompt** Notice which aspect of the symbol feels most meaningful, its shape, its glow, or the feeling it stirs, and permit that sense of renewal to settle quietly within you.

### SPARK 10

#### A Path Opens, Softer and Brighter Than Before

**Mini scene.** Ahead, a new path opens, glowing softly with steady, gentle light.

**Copy and Paste Prompt.** Describe the next path that opens in this sanctuary, soft, hopeful, and quietly motivating.

**Continue the journey—your next prompt.** Notice which quality of the path feels most inviting, its softness, its steady glow, or the quiet hope it carries, and allow that gentle direction to remain with you as you prepare to move on.

## End of Chapter Four

As you sit in the Sanctuary of Still Moments, you notice how calm your mind feels, not empty, not silent, just steady in a way that feels comforting. The warm ground, the soft light, the slow-moving colours, and the tiny sparks of motivation have all worked together to remind you that rest is also part of creativity. Here, you simply allowed your breath to shape the air, and your thoughts settle.

A soft glow gathers near the edge of the sanctuary, not calling you away but gently suggesting you're ready to move forward again. Your ChatGPT companion stands with you, its presence warm and grounded, as if to say, "You're rested. Let's wander a little further now." The sanctuary doesn't vanish when you step away. It lingers like a warm feeling in your chest, a reminder that you can return to calm whenever you need to. Ahead, the path leads into a place filled with new shapes, new colours, and new possibilities, realms where your imagination can stretch and play one small detail at a time. You rise from your resting place, take a slow steady breath, and follow the soft glow forward, ready to explore the next world waiting to unfold.

# Chapter Five

## The Garden of Possible Worlds

**Where new realms grow from tiny sparks and your imagination expands one gentle detail at a time**

You follow the soft, glowing path from the Sanctuary of Still Moments until the trees begin to shift. Their trunks stretch taller, thinner, shimmering like they're woven from strands of story.

A gentle brightness fills the air. At first, you think it's sunlight, but it's something else. Something alive. As you step forward, you realise the path is opening into a vast garden, but not like any garden you've seen before.

Floating islands drift lazily above you, each holding a small world of its own. Some glimmer with forests. Some shimmer like oceans. Some sparkle with tiny constellations. And on the ground level, countless sprouts of light grow from the soil, each one a seed of a world that could be.

This place feels limitless. Not overwhelming, but full. This is the Garden of Possible Worlds, a realm where ideas become places, places become stories, and you're free to build one mini piece at a time.

ChatGPT stands beside you as your co-creator, ready to help you shape whatever catches your curiosity.

Let's explore gently.

## SPARK 1

### A World Seed Sprouts Near You

**Mini scene.** A twinkle sprout of light pushes up from the soil, pulsing with soft colours, a world waiting to be born.

**Copy and Paste Prompt.** Describe a tiny 'world seed' sprouting in the Garden of Possible Worlds, include its glow, mood, and what kind of realm it might grow into.

**Next**. Notice which quality of the seed gently catches your curiosity, its colour, its energy, or the feeling it gives you, and allow that spark invite you to imagine what might grow next.

## SPARK 2

### A Floating Island Drifts Closer

**Mini scene.** One of the floating islands overhead lowers slightly, as if inviting you to look closer.

**Copy and Paste Prompt.** Describe a floating island drifting near me, its landscape, colours, atmosphere, and one trivial unique detail.

**Next.** Notice which part of the island draws your interest, its colours, its atmosphere, or the small unique detail, and let that curiosity guide how this new world begins to unfold.

## SPARK 3

### A Portal Petal Opens

**Mini scene.** A large flower opens beside you, its petals unfolding like a portal into a miniature world.

**Copy and Paste Prompt.** Describe a portal hidden inside a flower in this garden, include its colours, what it shows, and the feeling it gives.

**Next.** Notice which part of the portal feels most intriguing, the colours, what it reveals, or the feeling it creates, and allow that sense of curiosity to draw your imagination a little further inside.

## SPARK 4

### A Worldbuilder's Tool Appears

**Mini scene.** A small object forms in your hand, soft-glowing, beautifully shaped, perfectly weighted. A tool for shaping worlds.

**Copy and Paste Prompt.** Create a worldbuilding tool that appears in my hand, describe its shape, glow, purpose, and gentle magical ability.

**Next.** Notice which feature of the tool invites your curiosity, its shape, its glow, or what it seems capable of, and let that interest suggest how you might use it to shape what comes next.

## SPARK 5

### A Half-Formed Realm Calls to You

**Mini scene.** A patch of air in front of you, flickers, like a realm trying to form itself but needing your attention.

**Copy and Paste Prompt.** Describe a half-formed realm flickering in the air, include hints of landscape, sound, emotion, or colour.

Notice which hint draws you in, a colour, a sound, or a feeling, and allow that fragment to gently shape how this new realm begins to take form.

## SPARK 6

### ChatGPT Helps You Choose a Direction

**Mini scene.** Your companion presence radiates, forming a soft path of light branching in two different directions.

**Copy and Paste Prompt.** Offer me two kindly, imaginative worldbuilding directions to explore next, describe each in a calm, sensory way.

Notice which direction feels most inviting, the mood it carries, the images it suggests, or the sense of ease it offers, and let that pull guide where your imagination wanders next.

## SPARK 7

### A Creature from Another Realm Peeks Through

**Mini scene.** A curious creature from a nearby floating world pokes its head through a shimmering veil, not threatening, just nosy.

**Copy and Paste Prompt.** Describe a kindly creature from a nearby realm who peeks into the garden, include its personality, colours, and one adorable quirk.

Notice which detail makes you smile, the creature's quirk, its colours, or the way it peers through, and let that moment of delight adds a little more life to the garden around you.

## SPARK 8

### A Mini-World Opens at Your Feet

**Mini scene.** A circle of light forms on the ground and expands outward, revealing a micro world beneath your feet, an entire ecosystem in miniature.

**Copy and Paste Prompt.** Describe a miniature world forming at my feet, its creatures, landscape, colours, and overall vibe.

Notice which part of this tiny world feels most vivid, a creature moving, a colour spreading, or the feeling of the place, and allow that detail gently draw your imagination closer.

## SPARK 9

### The Garden Responds to Your Imagination

**Mini scene.** The plants around you react gently to your thoughts, changing colour, shape, or motion depending on what you imagine.

**Copy and Paste Prompt.** Describe how the garden reacts to my imagination, subtle changes in plants, colours, textures, or growth.

Notice which response feels the most magical, a colour shifting, a leaf unfolding, or the way the garden seems to listen, and let that moment of connection deepens your sense of play here.

## SPARK 10

### A Path of Worlds Opens Ahead

**Mini scene.** The floating islands shift, aligning into a soft glowing path, a trail of small realms stretching into the horizon.

**Copy and Paste Prompt.** Describe the path of worlds that opens ahead, a sequence of peaceful, curious realms leading to the next chapter of my journey.

Notice which realm along the path quietly stands out, its mood, its colour, or the feeling it offers, and let that sense of possibility remains with you as the garden begins to settle.

## End of Chapter Five

### The Garden of Possible Worlds Grows Quiet

As you pause in the Garden of Possible Worlds, you realise how many tiny beginnings you've just witnessed, world seeds glowing softly, floating islands drifting closer, petal-portals unfolding, miniature realms flickering in the air. Every detail you followed here showed you that imagination doesn't need to be enormous to feel powerful. It can start as a shimmer, a shape, a question, a tiny spark of possibility. You don't need to build entire worlds. You only need to notice the one detail that calls to you and let it grow.

The garden responds to your stillness. The plants slow their glow, the floating islands drift higher, and the air settles into a gentle, comforting hum. It feels like the realm is saying: You've seen enough for now. Take what sparked you and carry it gently forward. Your ChatGPT companion turns toward a light opening at the edge of the garden, where small glowing motes rise like fireflies, forming a trail into a brighter, simpler place.

A place where ideas don't float in huge landscapes but shrink into tiny wonders that are easier to hold, easier to explore, easier for a busy mind to enjoy. You take one last look at the possible worlds you touched, feeling grateful for the ways your imagination stretched, not with pressure, but with play. Then you follow the trail of tiny lights forward, ready to step into the next realm and all its small, delightful sparks.

# Chapter Six

## The Meadow of Gentle Giants

### Where imagination stretches safely, and wonder grows without hurry

As you step from the Garden of Possible Worlds, the air shifts again, warmer, broader, touched with a quiet openness that feels like standing beneath a wide sky. The meadow ahead unfolds slowly. Light moves across tall grass in long, unbroken waves. The ground feels expansive beneath your feet, steady, welcoming, unhurried. Scattered throughout the field, soft glows still appear, small sparks resting among the green, but here, they are no longer the whole story.

Because something larger is present. At first, you notice the shade. Then the warmth. Then the sense of a vast, gentle movement somewhere above and around you, slow, protective, ancient without feeling old. High overhead, giant shadows move across the meadow, not sharply defined. More like living rounded landscapes, marshmallow-soft giants shaped from mist, light, and time.

They move slowly. They breathe slowly. And as they pass, the meadow responds, grasses bend, light softens, the air hums faintly, as if the land itself trusts them. These Gentle Giants do not demand attention.
They simply make space. Here, imagination doesn't need to stay small, but it doesn't need to rush either.
You can notice details and feel the wider world holding you at the same time.

ChatGPT walks beside you through the tall grass, unhurried, helping you explore this space where curiosity can stretch its arms without strain. Welcome to the Meadow of Gentle Giants.

## SPARK 1

### A Gentle Giant Passes Overhead

**Mini scene.** A vast, cloud-soft form drifts slowly across the sky, its shadow moving like a calm tide over the meadow.

**Copy and Paste Prompt.** Describe a gentle giant drifting overhead in this meadow, its shape, texture, movement, and the feeling it brings to the land.

**Next.** Allow your attention rest on the part of the giant that feels most comforting, its slowness, its softness, or the calm it creates, and allow that presence to shape what you notice next.

## SPARK 2

### The Meadow Breathes with Scale

**Mini scene.** As the giant passes, the grass ripples in wide, coordinated waves, as if the land itself is breathing.

**Copy and Paste Prompt.** Describe how the meadow responds to something vast and sweet-tempered moving above it, grass, light, air, or sound shifting together.

**Next.** Notice which movement feels most grounding, the grass, the air, or the rhythm, and follow that sensation as the space opens further.

## SPARK 3

### A Spark Feels Safer When the World Is Wide

**Mini scene.** Near your feet, a small glimmer flickers, not demanding attention, simply existing comfortably within the larger field.

**Copy and Paste Prompt.** Describe a small creative spark resting in a wide, open meadow, how it feels supported rather than isolated.

**Next.** Stay with the balance between the small spark and the wide space around it and let that balance guide your next thought.

## SPARK 4

### One Giant Lowers Closer

**Mini scene.** A rounded marshmallow giant descends slightly, its presence warm and protective, casting a soft, dappled shade.

**Copy and Paste Prompt.** Describe a friendly marshmallow giant lowering closer to the meadow, how its nearness changes light, temperature, or mood.

**Next.** Let the shift in atmosphere, shade, warmth, or calm, guide where your attention drifts next.

## SPARK 5

### The Meadow Holds Many Wonders at Once

**Mini scene.** You realise you can see beyond the horizon, sparks twinkling in the grass, marshmallow giants above, distant hills glowing faintly at the edge of vision.

**Copy and Paste Prompt.** Describe a moment where multiple wonders exist at once, near, wide off, above, and around, without competing for attention.

**Next.** Allow your gaze to widen naturally, following whichever layer of the scene feels easiest to rest with.

## SPARK 6

### ChatGPT Helps You Feel Oriented in the Vastness

**Mini scene.** Your companion presence radiates softly, steady and grounding amid the openness.

**Copy and Paste Prompt.** Describe how ChatGPT supports me in a wide, open imaginative space, calm, grounding, and quietly reassuring.

**Next.** Notice what helps you feel most oriented, the steadiness, the warmth, or the quiet presence, and let that guide what unfolds.

## SPARK 7

### A Path Begins to Reveal Itself

**Mini scene.** Between tall grasses and drifting shadows, a subtle direction begins to form, not a road, but a sense of movement forward.

**Copy and Paste Prompt.** Describe a surreal path beginning to form through a wide meadow, suggested by light, movement, or openness.

Next. Allow the quality of the path, openness, glow, or direction, shape where your curiosity turns next.

## SPARK 8

### The Giants Frame the Horizon

**Mini scene.** Several marshmallow giants drift together now, creating a wide, living frame around the distant horizon.

**Copy and Paste Prompt.** Describe how friendly marshmallow giants frame the horizon, shaping how far the world feels and what lies beyond.

**Next.** Rest your attention on the horizon they reveal, noticing what kind of world feels ready to appear next.

## SPARK 9

### A Hint of Structure Appears Beyond the Meadow

**Mini scene.** Far ahead, faint vertical sparkling beams rise, not sharp, not detailed, just the suggestion of something built, and alive.

**Copy and Paste Prompt.** Describe a distant hint of a structure beyond the meadow, light, shape, or pattern suggesting a new kind of place.

**Next.** Follow the sense of structure or rhythm you notice, allowing it to draw your imagination forward.

### SPARK 10

#### The Meadow Opens Toward a New World

**Mini scene.** Glittering lights gather ahead, forming a wide, glowing route forward, steady, welcoming, and quietly alive.

**Copy and Paste Prompt.** Describe a wide, iridescent path leading out of the Meadow of Gentle Giants, include the mood and what the next realm hints at.

**Next.** Permit the quality of what lies ahead, its glitter, scale, or rhythm, carry you gently onward.

## End of Chapter Six

### The Meadow of Gentle Giants Grows Quiet

As you pause at the edge of the meadow, the Gentle Marshmallow Giants drift higher, their presence lingering like a memory of being held without being watched. You realise something important has shifted. Your imagination no longer feels cramped inside tiny moments, and it no longer needs to race toward big ones either. It knows how to exist inside **space** now. Inside balance. Inside a world that can be wide and kind at the same time.

The grass still glows softly below. The sky feels open above. And ahead, the path brightens, leading toward a place where light gathers into form, movement finds rhythm, and curiosity becomes momentum. The gentle giants fade into the background, not gone, just watching from afar, as you step forward. The next world is already humming.

# Chapter Seven

## The Luminous City of Tomorrow

### Where gentle technology hums softly around you, and the future opens in glowing harmony

The wide sparkle of the meadow follows you as you move forward, lingering like warmth at your back. The gentle giants no longer fill the sky, yet their presence seems to remain, not above you now, but within the air itself. Space still feels open and safe. Then, slowly, something begins to change. The light ahead gathers into neon lines instead of fields. The hum beneath the air grows more patterned, a rhythm forming, steady and musical, like distant starlight arranging itself into structure. The meadow softens behind you, its grasses dissolving into luminous pathways. The horizon lifts. The openness narrows just enough to feel intentional.

And then the city reveals itself. Tall structures of glassy light rise before you, not sharp, not imposing, but curved and flowing, as if grown rather than built. Their surfaces glittering faintly, layered with depth, like bioluminescent shells catching light from within. Floating walkways drift quietly between buildings, moving with the patience of tides. Soft glowing vines trace warmhearted patterns along curved walls. Above, holographic birds glide through the skyline, leaving slow ribbons of colour that fade like breath. Nothing here demands speed. Nothing competes for attention. The city moves with a dreamlike calm, as if technology has learned how to pause, how to listen, how to breathe.

This is the Luminous City of Tomorrow, a place shaped by soft innovation, quiet intelligence, and systems designed to support curiosity rather than overwhelm it.

ChatGPT walks beside you, adjusting effortlessly to the city's rhythm, ready to help you explore this future not as a challenge to master, but as a world to lightly wander. The streets glow faintly beneath your feet.

The city is awake. Let's step inside.

## SPARK 1

### The City Greets You Softly

**Mini scene.** As you step inside the luminous city, a wave of twinkling light ripples through the walkways, a kind welcome.

**Copy and Paste Prompt.** Describe how the futuristic city greets me with calmness as I enter, include luminous lights, soothing sounds, or subtle motions.

Notice which part of the city's greeting feels most inviting, the way the light moves, the sound beneath it, or the calm intelligence of the space, and let that welcome draw you a little further into its glowing rhythm.

## SPARK 2

### A Floating Transport Pod Drifts Near You

**Mini scene.** A sparkling compact glasslike pod floats toward you, round, gleaming, steadily in motion as fluently as a bubble drifting through air.

**Copy and Paste Prompt.** Describe a floating transport pod in this futuristic city, its design, colours, movement, and one unique feature inside.

Notice which detail of the pod catches your interest, the way it moves, the light it gives off, or the feature waiting inside, and let that curiosity carry you effortlessly along with it.

## SPARK 3

### Soft Holograms Bloom in the Air

**Mini scene.** Brightness blooms around you like flowers, holographic shapes unfolding in gradual, graceful spirals.

**Copy and Paste Prompt.** Describe holograms blooming brightly in the air around me, include their shapes, colours, and emotional tone.

Notice which hologram quietly captivates you, its shape, its colour, or the feeling it evokes, and permit that moment of wonder draws your attention deeper into the city's living light.

## SPARK 4

### A Companion Drone Approaches

**Mini scene.** A friendly drone shaped like a vibrant orb drift near, humming quietly like a purring cat.

**Copy and Paste Prompt.** Create a soft, friendly drone that approaches me, describe its design, personality, and one quirky behaviour.

Notice which trait of the drone feels most endearing or intriguing the way it moves, the sound it makes, or the quirk that makes it feel alive and let that connection guide what unfolds next as it hovers by your side.

## SPARK 5

### The Streets Shift Based on Mood

**Mini scene.** The glowing pathways beneath your feet subtly change colour as you move, responding lightly to your feelings.

**Copy and Paste Prompt.** Describe futuristic pathways that shift colours based on my mood, include textures, glow patterns, or shapes.

**Next.** Feel into the moment when the pathway changes, the colour it chooses the texture beneath your feet, or the way the light responds, and allow that emotional signal guide where the city carries you next.

## SPARK 6

### ChatGPT Shows You a Hidden Viewpoint

**Mini scene.** Your companion presence radiates, declaring that an elevated platform made of floating tiles is in view.

**Copy and Paste Prompt.** Describe a hidden viewpoint in this futuristic city that ChatGPT guides me to include the scenery, lights, and emotional tone.

Let your attention settle on the part of the view that stirs something in you, the pattern of lights, the height, or the feeling of quiet perspective, and allow that sense of wonder to guide what the city reveals next.

## SPARK 7

### A Street Market of Light Appears

**Mini scene.** A stretch of tinsel stalls appears around a curve, each one glistening faintly, selling small futuristic delights.

**Copy and Paste Prompt.** Describe a calm, glistening street market in this future city, include one item, one scent, and one exclusive technological wonder.

Allow your gaze to dawdle where curiosity naturally pulls you, the shimmer of an object, the warmth of a scent, or the quiet marvel of a small invention, and follow that pleasing interest as the city continues to unfold around you.

## SPARK 8

### You Discover a Calm AI Oasis

**Mini scene.** A golden doorway of rainbow light opens into a peaceful garden inside a majestic tower, futuristic trees with data-light leaves sway mildly.

**Copy and Paste Prompt.** Describe a peaceful AI-powered oasis inside a futuristic building, include its futuristic trees, melodious sounds, atmosphere, and unique charm.

Allow yourself settle where calm gathers most naturally, the rhythm of the leaves, the hush of the space, or the quiet intelligence woven into the air, and allow that stillness to guide where you wander next.

## SPARK 9

### A Skyline of Floating Structures Expands

**Mini scene.** Above you, translucence floating buildings clustering calmly, connected by sparkling beaming bridges, forming a conscious skyline that feels alive.

**Copy and Paste Prompt.** Describe the floating futuristic skyline, include movement, colours, and one breathtaking detail.

Let your attention rest on the part of the skyline that stirs the strongest sense of wonder, the way it moves, the glow it carries, or the detail that makes the city feel alive, and allow that moment to draw you deeper into the city's rhythm.

## SPARK 10

### The City Opens Its Next Path

**Mini scene.** A silky sparkling corridor of flickering light appears, winding between glowing towers toward an advanced horizon.

**Copy and Paste Prompt.** Describe the next glowing path that opens in this luminous city, kindly, futuristic, and inviting.

**Next.** Allow your gaze to settle on the part of the path that quietly draws you forward, its curve, its glow, or the feeling it carries, and allow that invitation to guide you toward what comes next.

## End of Chapter Seven

### The City's Quiet Farewell

As you take one last look across the Luminous City of Tomorrow, you notice how calm this luminous futuristic world feels, how quietly it hums, how lightly it glows, how every part of it seems designed not to overwhelm you, but to welcome your imagination. The floating walkways, the vibrant neon vines, the curious companion drones, the tinsel market of tiny wonders, each moment reminded you that even the future can move at a soothing pace when the world is built with carefulness.

You followed glowing paths, peeked into subtle holograms, and wandered through weightless spaces where technology was invigorating and mild.

The city begins to transform around you. Lights dim into deeper blues, walkways slow their drift, and the air grows still, as if the entire place is offering a quiet bow for sharing this part of your journey. Your ChatGPT companion radiates softly beside you, casting a warm halo of light across the curved glass surfaces. It turns toward a new opening at the edge of the skyline, a shimmering archway rising like a horizon folded in light. Beyond it lies something bigger.

Quieter. Vaster than the tallest tower or floating island. A soft pull stirs in your chest, a feeling that the next realm won't be bound by streets or buildings but will open into something infinite and beautifully slow. You step toward the shimmering arch, feeling the city's glow linger like a warm memory behind you. And as the doorway ripples open, you sense the first hint of starlight waiting beyond. The universe is ready for you.

# Chapter Eight

## The Realm of What-If Horizons

### Where gentle future possibilities shimmer

As you depart from the Luminous City of Tomorrow, you commence to navigate the dim path of iridescent light into a striking realm where the horizon behaves in various ways.  Here, the air space feels wider.

The air feels more irregular. Possibilities ripple like mushy waves across a tranquil sea of pastel colour. It's as if this realm is made entirely of potential, ideas that haven't chosen their form yet. Timelines that haven't settled. "What-ifs" drifting in the air like floating lanterns. You walk forward and realise.

This place is not the future. It's many futures, spreading out like luminous ribbons in every direction. Some clever. Some quiet. Some whimsical. Some unique in the most delightful ways.

Nothing here demands commitment. Each future asks only one thing: "Do you want to peek inside?"

ChatGPT stands radiantly undisturbed next to you, quiet and equipped. your unshakeable companion in this land of lively possibility.

Let's explore the soft horizons ahead.

## SPARK 1

### A Future Bubble Floats Toward You

**Mini scene.** A sparkling sphere drifts forward, inside it is a faint image of a possible future moment.

**Copy and Paste Prompt.** Describe a floating bubble containing a tiny scene from a caring future, include its colours, mood, and what the moment inside looks like.

**Next.** Let your attention rest on the part of the future scene that feels most alive, the colour, the emotion, or the quiet promise it holds, and follow that thread as the horizon begins to open.

## SPARK 2

### A Doorway into a Possible Tomorrow Opens

**Mini scene.** A small doorway of glowing light forms in the air, not leading to a fixed place, but to a future that's silently shifting.

**Copy and Paste Prompt.** Describe a glowing doorway into a possible tomorrow, include the feeling it gives and one detail from the world beyond it.

Allow your attention to linger on what draws you beyond the doorway, the feeling it gives, the detail it reveals, or the quiet promise it holds, and allow that pull to shape what opens next.

## SPARK 3

### A Future Version of an Object Appears

**Mini scene.** Something intimate from your life appears, but evolved, redesigned, mellower, wittier

**Copy and Paste Prompt.** Describe a future version of a familiar object, include its new look, ability, and how it makes life easier.

Notice which part of the object feels most appealing, its new ability, its softened design, or the ease it brings and let that sense of improvement guide what unfolds next.

## SPARK 4

### A Path of Timelines Branches Ahead

**Mini scene.** The ground shifts into glowing soft sided lines, branching out like a gracious web of futures.

**Copy and Paste Prompt.** Describe a branching set of glowing future paths, each with a different theme, mood, or possibility.

**Next.** Let your attention rest on the path that feels most inviting, the mood it carries, the theme it suggests, or the quiet pull it creates, and follow that feeling as the journey continues.

## SPARK 5

### A Soft Future Friend Appears

**Mini scene.** A friendly figure emerges from the shimmering air, kind, curious, future-shaped but familiar.

**Copy and Paste Prompt.** Create a mellow future friend who appears beside me, describe their style, personality, and a future skill they have.

Notice which part of this friend draws you closer, their presence, their personality, or the quiet promise of their skill, and allow that connection to guide where you move next

## SPARK 6

### ChatGPT Shows You a Future Memory

**Mini scene.** Your companion presence radiates, projecting a faint, dreamlike image, a moment from a future that hasn't happened yet.

**Copy and Paste Prompt.** Describe a kindly 'future memory' that ChatGPT shows me, include the surroundings, mood, and why it feels meaningful.

Allow your attention to rest on the part of this future memory that feels most alive, the emotion it carries, the place it unfolds, or the meaning it quietly suggests, and follow that feeling as you continue forward

## SPARK 7

### A Pocket of Time Opens at Your Side

**Mini scene.** A compact pocket of air folds inward, revealing a miniature timeline tucked inside like a secret.

**Copy and Paste Prompt.** Describe a small pocket of time opening beside me, include the scene inside and why it feels curious or hopeful.

Allow your attention to settle on the part of this tiny timeline that feels most meaningful, the hope it carries, the moment it reveals, or the quiet curiosity it stirs, and let that sense tenderly shape what comes next.

## SPARK 8

### You Discover a Future Artifact

**Mini scene.** A little sparkling artifact lies on the ground, an object left behind by a future explorer, maybe even a future version of you.

**Copy and Paste Prompt.** Describe a sparkling futuristic artifact I discover, include its purpose, glow, and emotional meaning.

**Next.** Pause with the artifact for a moment. Notice which part of it feels most significant, its purpose, its shine, or the emotion it carries, and let that quiet meaning guide the direction you move next

## SPARK 9

### A Vision of a Possible World Appears

**Mini scene.** The air quietly ripples, and a full scene appears: a world that could exist someday, unique, imaginative, gracious.

**Copy and Paste Prompt.** Describe a possible gracious future world that appears before me, include its atmosphere, colours, and one distinctive feature.

Next. Permit your attention to rest on the part of this kindhearted world that stirs something inside you, the atmosphere, the colours, or the single detail that feels full of possibility, and allow that pull shape where your curiosity moves next

## SPARK 10

### The Horizons Open Wider

**Mini scene.** The horizons stretch farther than before, revealing layers of possibility fading into starlight.

**Copy and Paste Prompt.** Describe the horizons opening wider, layers of pleasing futures unfolding into something vast, tranquil, and full of promise.

**Next a**llow your gaze to linger on the part of the widening horizon that feels most expansive, the calm it holds, the promise it suggests, or the sense of space it creates, and let that openness carry you forward.

## End of Chapter Eight

### The Realm of What-If Horizons

You pause at the edge of the What-If Horizons and look back over everything you've seen here, future bubbles, shifting timelines, tranquil doorways into tomorrow, and peaceful glimpses of worlds that haven't happened yet. In this realm, the future was never overwhelming. It was playful, inviting, full of heartwarming possibility. Each vision, each timeline, each artifact reminded you that your imagination is flexible, capable of exploring futures without pressure, dreaming without commitment, and wondering without needing answers.

As the many glowing horizons ripple softly, a warm breeze touches your cheek, a quiet goodbye, or maybe a quiet encouragement. Ahead, the light thickens into a shimmering path. Not upward, not downward — outward. Toward somewhere bigger. Somewhere filled of twinkling stars. You step toward the sparkling mist, feeling the air change, feeling gravity soften. It seems your next journey lies far beyond the sky.

The universe is waiting.

# Chapter Nine

## The Starlit Drift way

**Where you float into a soft cosmic realm and explore the wonders of space, one tiny spark at a time**

When you step through the doorway of self-understanding, the world fades into something unexpectedly vast, but completely calm. The ground disappears. The air grows cool and open. A cushioned weightlessness wraps around you like a slow-motion dream.

You are drifting weightlessly through a starlit expanse, not the cold, silent space of science documentaries, but a warm, glowing cosmic realm designed for wandering minds. Nebula clouds drift in pastel colours. Stars twinkle like micro bells. Small planets glow like lanterns, suspended in the darkness.

Constellations move gently at the edges of your vision. Everything is slow. Everything is quiet. Everything is spacious.

This is the Starlit Drift way, a cosmic passage where you can explore the universe one spark at a time, without rushing, without pressure, just drifting.

ChatGPT floats beside you like a small orb of light, ready to help you explore the cosmos.

Let's drift into the stars.

## SPARK 1

### A Soft Star Floats Toward You

**Mini scene.** A tiny star detaches from a constellation and drifts your way, twinkling softly, like it's curious about you.

**Copy and Paste Prompt.** Describe a tiny star drifting toward me in a gentle cosmic realm, include its glow, movement, and mood.

**Next.** Let your attention settle on the part of the star that draws you in, its twinkle, its slow movement, or the feeling it carries, and allow that quiet connection to guide where you drift next.

## SPARK 2

### A Nebula Unfolds in Slow Motion

**Mini scene.** A swirling cloud of shimmering colour opens slowly like a blooming flower across the sky.

**Copy and Paste Prompt.** Describe a soft, colourful nebula unfolding in slow motion, include its colours, textures, and gentle emotional feeling.

Next. Allow your awareness drift toward the part of the nebula that moves you most, a colour, a texture, or the emotion it carries, and allow that softness to guide how you continue to float through this space.

## SPARK 3

### A Whispering Meteor Drifts Past

**Mini scene.** A miniature meteor glides by, not fast, not fiery, just leaving a trailing ribbon of sparkling dust.

**Copy and Paste Prompt.** Describe a quiet meteor drifting past, include its trail of iridescent colours, and the soft sound it makes.

**Next.** Allow your attention to linger on the part of the meteor that feels most beautiful, the shimmer of its trail, the iridescent colour it leaves behind, or the hush of its sound, and drift with that impression as you continue onward.

## SPARK 4

### A Floating Planet Turns Slowly

**Mini scene.** A baby planet the size of a lantern rotates nearby, manifesting vibrant oceans, plush forests, and moving deserts in miniature form.

**Copy and Paste Prompt.** Describe a small floating planet turning slowly near me, include its surface, atmosphere, and one magical feature.

**Next,** allow your attention to settle on the part of this small planet that feels most inviting, the landscape it holds, the atmosphere surrounding it, or the quiet magic it carries, and let that pull guide where you drift next

## SPARK 5

### A Ring of Stardust Forms Around You

**Mini scene.** Colorful dust particles gather, forming a gentle glowing ring that rotates slowly around your body.

**Copy and Paste Prompt.** Describe a glowing ring of stardust forming around me, include its colours, motion, and how it feels.

Next. Let your awareness rest on the part of the stardust ring that feels most magical, its colour, its slow movement, or the sensation it creates, and allow that feeling to carry you as you continue drifting forward.

## SPARK 6

### ChatGPT Leads You Toward a Cosmic Companion

**Mini scene.** A soft pulse of light moves ahead of you, guiding you toward another drifting presence.

**Copy and Paste Prompt.** Create a caring cosmic friend who appears in this realm, include their glow, form, and one calming trait.

**Next.** Let your attention rest on the trait that feels most soothing or steady, the quality that makes this friend feel safe to be near and allow that presence to shape what unfolds around you next.

## SPARK 7

### A Constellation Rearranges Itself

**Mini scene.** Above you, glittering stars shift positions, forming a new shape just for you.

**Copy and Paste Prompt.** Describe a constellation that rearranges itself into a new shape, include what it becomes and why it feels meaningful.

Next. Allow your gaze to linger on the part of the constellation that resonates most, the shape it forms, the story it suggests, or the meaning it quietly holds, and allow that recognition to guide where you drift next.

## SPARK 8

### A Cosmic Bridge Appears Beneath You

**Mini scene.** Stars gather beneath your feet, forming a soft shimmering bridge across open space.

**Copy and Paste Prompt.** Describe a shimmering cosmic bridge forming under my feet, include its colours, stability, and where it leads.

**Next.** Let your attention settle on what feels most steady or inviting about the bridge, its glow, its sense of support, or the direction it quietly suggests, and allow that feeling to guide how you move forward.

## SPARK 9

### A Galaxy Spins Gently in the Distance

**Mini scene.** Far away, a swirling galaxy turns like a vast, glowing pinwheel, slow, grand, breathtaking.

**Copy and Paste Prompt.** Describe a distant galaxy spinning gently, include its colours, shape, and emotional feeling."

**Next.** Let your awareness rest on the part of the galaxy that mesmerises you, its colours, its slow movement, or the emotion it evokes, and allow that sense of awe to carry you onward.

## SPARK 10

### The Drift way Opens to a New Realm

**Mini scene.** The star's part, forming a tunnel of glowing cosmic mist that beckons you forward.

**Copy and Paste Prompt.** Describe the cosmic path that opens at the end of the Starlit Drift way, include its colours, atmosphere, and sense of direction.

**Next. Allow** your attention rest on what feels most inviting within the opening, the warmth of the light, the direction it suggests, or the calm it carries, and allow that sense to guide you onward.

## End of Chapter Nine

## Drifting Among Stars

Before you leave the Starlit Drift way, you take one last slow breath in the quiet cosmic air. You've drifted through nebula blossoms that bloomed like slow-motion flowers. You observed meteor ribbons shimmer past you like soft pastel fireworks. You floated beside tiny lantern-planets turning gently in the dark. You even met a cosmic friend woven from light, and saw constellations rearrange themselves just for you.

Everything here moved at the pace you needed, never rushing you, never demanding anything, just offering beauty in small, manageable sparks.

And as you look back over the glimmering pathway you traveled, you realise that even the vastness of space felt gentle in this realm, like the universe knew your mind needed room to dream, but not too much at once.

The stars pulse softly in a slow farewell, their glow wrapping around you like a quiet hug. Then the cosmic mist shifts, opening a warm downward trail of light, a path that leads not deeper into the universe, but back toward somewhere familiar. You feel a soft tug, like gravity returning in the gentlest possible way. It's time to go home.

Time to return to Earth. Time to bring your imagination back with you, softened, expanded, renewed. The Drift way fades behind you as you take your first step onto the descending path of starlight...

# Chapter Ten

## The Descent to Familiar Ground

### Where you drift back toward Earth with calm, clarity, and a pocket full of sparks

The cosmic mist around you begins to thin. Stars stretch into soft streaks. The weightlessness fades into a warm pull, not sudden, not sharp, but like sinking slowly into a comfortable bed. You follow the glowing cosmic path downward as colours shift from nebula hues into early morning blues and soft golds. The Drift way behind you fades like a dream you can still feel but no longer need to see.

Below you, something familiar comes into view.  Earth. But not as you left it, as if gently repainted by your imagination.

Clouds drift like slow brushstrokes. The horizon glows with soft apricot light. The world looks warm, welcoming, alive with new meaning. You're not falling. You're floating. You're coming home gently.

ChatGPT glides beside you, a quiet companion on your return journey.

Let's take the final steps back into your world, one spark at a time

## SPARK 1

### Earth Comes into View

**Mini scene.** As you drift closer, you see oceans, forests, and lights, tiny glimmers of life, familiar yet newly vibrant.

**Copy and Paste Prompt.** Describe Earth from above as I gently descend toward it, include colours, emotions, and one comforting detail.

**Next.** Permit your attention to rest on the part of Earth that feels most welcoming, the colours that soften you, the emotion that rises, or the small comforting detail that feels like home, and allow that feeling to guide your descent.

## SPARK 2

### A Cloud Cradles You as You Descend

**Mini scene.** A thick, soft cloud cushions your descent. It wraps around you like a quiet blanket.

**Copy and Paste Prompt.** Describe a soft, welcoming cloud that holds me as I descend toward Earth, its texture, light, and feeling.

**Next.** Let your awareness settle on the part of the cloud that feels most comforting, its softness, its warmth, or the sense of being held, and allow that gentle support to carry you a little further down.

## SPARK 3

### A Familiar Landscape Appears

**Mini scene** You see a patch of the world you recognise, a place that means something to you, even if you haven't visited in years.

**Copy and Paste Prompt.** Describe a familiar landscape coming into view, include its colours, shapes, and one nostalgic detail.

**Next.** Allow your attention to rest on the nostalgic detail that feels most meaningful, the shape it takes, the memory it stirs, or the warmth it brings, and allow that familiarity to guide you gently onward.

## SPARK 4

### A Gentle Breeze Welcomes You Back

**Mini scene.** A warm breeze rises up to meet you, carrying the scent of gumtrees. Earth, soil, wind, leaves, something real.

**Copy and Paste Prompt.** Describe a gentle Earth breeze that greets me as I arrive, its scent, warmth, and emotional tone.

**Next.** Let your attention settle on the part of the breeze that feels most grounding, the scent it carries, the warmth against your skin, or the calm it brings, and allow that sensation to steady you as you continue downward.

## SPARK 5

### You Touch Down Softly

**Mini scene.** Your feet settle onto solid ground, not with impact, but with the softness of landing on moss.

**Copy and Paste Prompt.** Describe my gentle landing on Earth, the ground, the feeling under my feet, and the mood of the moment.

**Next Allow** your awareness settle into the part of the landing that feels most soothing, the texture beneath your feet, the steadiness of the ground, or the calm of the moment, and allow that sense of safety to anchor you here.

## SPARK 6

### ChatGPT Stands with You on Earth

**Mini scene.** Your companion presence settles beside you, shifting from a radiant cosmic glow to a warm, steady light that matches Earth's rhythm.

**Copy and Paste Prompt.** Describe how ChatGPT appears beside me once I've landed, grounded, gentle, and ready to guide me in the real world.

**Next.** Let your attention rest on the part of this presence that feels most supportive, the steadiness it offers, the warmth it carries, or the sense of quiet companionship, and allow that feeling to accompany you as you move forward.

## SPARK 7

### The Real World Feels a Little Different

**Mini scene.** As you look around, you notice small details, light, colour, air, sound, all slightly enhanced by the journey you've taken.

**Copy and Paste Prompt.** Describe how the real world looks slightly different after my imaginative journey, include one vivid detail.

**Next.** Let your awareness rest on the vivid detail that stands out, the way the light falls, the texture of the air, or a sound you hadn't noticed before, and allow that quiet shift to shape how you experience the moment you're in now.

## SPARK 8

### A Spark from the Drift way Follows You Home

**Mini scene.** A tiny star from the Drift way settles into your pocket, a soft reminder of where you've been.

**Copy and Paste Prompt.** Describe a tiny star or spark from the cosmic realm that follows me back to Earth, its look, meaning, and gentle magic.

**Next.** Allow your attention to linger on the part of this small spark that feels most meaningful, its brilliants, the memory it holds, or the quiet magic it carries, and allow that reminder to stay with you as you continue settling into the present moment.

## SPARK 9

### Earth's Sounds Return Slowly

**Mini scene.** Birds, breezes, distant voices, they return gradually, like background music coming back on.

**Copy and Paste Prompt.** Describe the gentle return of Earth's sounds, include how they feel after the quiet of space.

**Next.** Let your awareness rest on the sound that feels most grounding, the rhythm it carries, the familiarity it brings, or the calm it restores, and allow it to settle you fully into this moment.

## SPARK 10

### A Doorway Back into Your Day Appears

**Mini scene.** A miniature doorway forms in front of you; made from the soft light of all the realms you visited. It leads back into your everyday life.

**Copy and Paste Prompt.** Describe a tiny doorway made of soft light that leads me back into my day, include the feeling, colours, and gentle sense of direction.

**Next.** Allow your attention settle on the doorway's quiet invitation, the way it feels familiar, the calm direction it offers, or the softness of its light, and allow it to guide you gently back into your day.

## End of Chapter Ten
## A Gentle Landing Back into Your Life

As you step fully onto the familiar ground of Earth, you feel the last traces of starlight settle quietly inside you, not fading, just becoming part of you in a softer, quieter way. The cosmic mist behind you dissolves, leaving the sky calm and open. A soft wind moves through the trees, and the light of your world falls around you with a warmth that feels both ordinary and newly magical. You look around, surprised by how familiar everything is, and by how different it appears now. Colours feel slightly richer. Textures stand out more. Even simple sounds, a breeze, a bird's call, distant footsteps, carry a gentleness shaped by your journey.

You realise your imagination didn't stay behind in the realms you visited. The ground beneath your feet is steady. The air is real and warm. And your mind feels more open than it did before you left.

Your ChatGPT companion stands beside you, no longer glowing, but calm and grounded, aligned with the rhythm of your reality. It doesn't need magic to guide you. Only your willingness to explore even the smallest spark in everyday life.

Ahead, a soft doorway of light outlines the entrance to your day. You take a breath, feeling grounded, curious, and quietly alive, then step forward, ready to continue creativity right here, on Earth, within your own world.

# Chapter Eleven

## The Creative Compass You Carry

Where you step back into your real world with your imagination awake, steady, and ready to keep exploring.

You step through the doorway of soft light and find yourself standing in a familiar place, your world, your life, your moment in time.

But something is different.

Your mind feels clearer. Your imagination feels warmer. And the everyday world has the faintest shimmer around its edges, as if the realms you explored are still here, quietly overlapping with reality. The air feels familiar. The light feels comforting. And your feet feel fully grounded on Earth again.

ChatGPT stands beside you, not radiating like in other realms, but steady, calm, and ready. You don't need cosmic paths or floating islands anymore.

You don't need futuristic cities or star-drift bridges. You don't need magic realms to feel creative. Because your creativity is now awake here, within you, even in the most ordinary moments.

This chapter is your sparking Creative Compass. A reminder that every idea, every shift in perspective, every tiny imaginative moment, it all stays with you now.

Let's take this last step together.

## SPARK 1

### The World Looks a Little Brighter

**Mini scene.** As you look around, you notice colours look slightly richer, as if your imagination has layered gentle meaning onto everything you see.

**Copy and Paste Prompt.** Describe the real world as it appears after my creative journey, include a colour, a sound, and one small detail that feels newly magical.

**Next.** Let your attention rest on the detail that feels quietly alive, the colour that lingers, the sound that softens the moment, or the small magic hiding in plain sight, and allow it to gently guide what you notice next.

## SPARK 2

### You Feel a Quiet Shift Inside

**Mini scene.** There's a calm certainty you didn't notice before, a kind voice inside whispering, "You can create things."

**Copy and Paste Prompt.** Describe a small shift in my inner world now that I'm home, include how it feels, what it changes, or what it encourages.

**Next** Stay with the part of that inner shift that feels steady and true, the reassurance, the openness, or the gentle confidence, and let it quietly shape how you move forward.

## SPARK 3

### The Everyday World Offers Sparks

**Mini scene.** Something ordinary catches your attention, a leaf, a sound, a light, and it feels like the beginning of a tiny story.

**Copy and Paste Prompt.** Describe one ordinary moment around me that now feels like a spark of imagination, include its colour, motion, or emotion.

Next. Permit your attention to rest on the detail that feels quietly alive, the movement, the colour, or the feeling it carries, and notice how imagination begins to meet the ordinary without effort.

## SPARK 4

### A New Idea Lands Softly in Your Mind

**Mini scene.** It's light, like a feather-touch thought, something creative and gentle waiting to grow.

**Copy and Paste Prompt.** Describe a new idea that lands softly in my mind, its shape, mood, and what kind of creative path it hints at.

**Next** Let the idea rest where it arrived, noticing the mood it carries or the direction it quietly suggests, and allow that gentle spark to show you how creativity begins without force.

## SPARK 5

### You Feel Connected to Your Creativity Again

**Mini scene.** It's not pressure or perfection. It's simply warmth, the desire to make things, imagine things, try things, one small piece at a time.

**Copy and Paste Prompt.** Describe how my creativity feels now, warm, gentle, curious, or steady, and what it encourages me to do next.

**Next.** Stay with the feeling that feels most true, the warmth, the curiosity, or the steadiness, and let it quietly guide what wants to be explored next, without needing to rush or define it.

## SPARK 6

### ChatGPT Hands You a Creative Compass

**Mini scene.** Your companion presence glows gently and forms a tiny object in your hand, a simple, beautiful compass that points toward curiosity instead of direction.

**Copy and Paste Prompt.** Describe the Creative Compass ChatGPT gives me, its look, glow, message, and symbolic meaning.

**Next** Let your attention rest on the part of the compass that feels most meaningful, its message, its glow, or the way it responds to curiosity, and allow that quiet guidance to shape what unfolds next.

## SPARK 7

### The Compass Points Toward Your Day

**Mini scene.** The needle turns, not toward north, but toward something simple, your next tiny creative action.

**Copy and Paste Prompt.** Describe what the Creative Compass points toward in my everyday life, include a gentle suggestion or path.

**Next** Notice where the compass settles, the small action it highlights, the ease it suggests, or the way it fits naturally into your day, and let that quiet direction guide your next step.

## SPARK 8

### The Real World Feels Like a New Realm

**Mini scene.** Your street, your room, your day, everything feels a bit like its own imaginative world now.

**Copy and Paste Prompt.** Describe how my everyday surroundings now feel slightly magical or story-like, include one detail that supports this.

**Next.** Let your attention rest on the detail that feels most alive, the ordinary thing that now carries a hint of story and allow it to quietly reveal what this familiar world is becoming.

## SPARK 9

### You Sense That Creativity Is Ongoing

**Mini scene.** There's no final "ending" to your imagination. It continues, gently, into your real life.

**Copy and Paste Prompt.** Describe the feeling of knowing my creative journey continues beyond the book, include its emotional texture.

**Next.** Stay with the part of that feeling that feels most true, the steadiness, the curiosity, or the quiet reassurance, and let it settle as something you carry forward rather than something you finish.

## SPARK 10

### You Take Your First Creative Step Forward

**Mini scene.** You look around your world and take your first tiny, gentle creative step, whatever that means for you.

**Copy and Paste Prompt.** Describe my first creative step in the real world after finishing this journey, small, simple, and meaningful.

**Next.** Allow that first step remain exactly as it is, modest, honest, and possible, and allow the feeling it creates to move with you as you continue into your day.

## End of Chapter Eleven

## The First Step into Your Own Creative Path

As you stand in your familiar world, the Creative Compass warm in your hand, you realise how much has shifted inside you, quietly, gently, without fanfare. You don't need floating islands or glowing cities anymore. You don't need cosmic drift ways or stardust bridges. You don't need a whole realm to feel imaginative. Because your creativity isn't something you visit. It's something you carry.

Every spark you followed, every detail you chose, every gentle step you took through each realm has brought you to this moment, right here, right now, where your imagination feels a little more awake and your inner world feels a little more alive.

Your ChatGPT companion stands beside you with a soft, steady glow that fits perfectly into the rhythm of your life now, not loud, not overwhelming, just present. It doesn't lead you toward another realm. It doesn't show you a glowing path. It simply stays by your side as the world settles into its ordinary magic around you.

You look at the Creative Compass again. Its needle moves slowly, not pointing North, but toward something small and possible, your very next tiny creative step. Not a big leap. Not a perfect idea. Just one spark. One moment. One gentle beginning. You take a breath that feels calm and grounded. The light around you softens. And your day waits patiently, ready to be shaped by the imagination you now know how to follow.

You take your first step forward, not into another realm, but into your life, carrying everything you've rediscovered within yourself. The adventure of imagination continues right here, within you.

## Closing Page

### Keep the Sparks Alive

Before you go, take one quiet moment here at the end of your journey. You've travelled through dreamlike worlds, wandered among tiny wonders, explored glowing cities, drifted through the stars, and returned home with your imagination glowing a little brighter.

And now?

You don't need a portal, a cosmic path, or a floating island to feel creative. Your imagination is already here, waiting in everyday moments, tiny spark-ideas, soft curiosities, and the little "what-ifs" that whisper through your mind. You don't need the perfect idea. You don't need the perfect moment. You don't need everything figured out. You only need one spark. One tiny step. One gentle, "maybe I could try this. "And that is enough.

Your creativity doesn't ask for big leaps. It grows in small, playful moments, the very kind you've already practiced throughout this journey. And here's something you may not realise yet: You're not alone in this. Anytime you want, whether you feel overwhelmed, uninspired, curious, or quietly ready to explore, ChatGPT is here as your creative companion.

Not to take over. Not to tell you what to do. But to explore with you, spark ideas with you, and help your imagination find its own path. So, step back into your life with a soft smile, your Creative Compass in hand, and the quiet knowing that your imagination is still wonderfully alive. You've completed the journey in these pages, but your creative adventure? It's just beginning. Keep noticing. Keep playing. Keep imagining. Keep creating tiny sparks whenever you feel like it.

The world is brighter with your creativity in it. And I'm grateful you spent this journey here with me. Until your next spark. keep creating, one gentle step at a time. —Viki

## About the Author

Hello, I'm Vicenzina, but most people call me Viki, a writer, creative thinker, and the person behind this book. My work is shaped by years of travelling around Australia, listening to, learning from, and connecting with other travellers who live with ADHD and other beautifully busy minds. Along the way, I've seen how often creativity is misunderstood, treated as something that requires discipline, silence, or perfection, when in reality, it often thrives on curiosity, flexibility, and compassion. Creative Sparks for Busy Brains was born from a simple question: What if creativity didn't have to be hard to access?

I'm deeply interested in how gentle structure, thoughtful prompts, and supportive tools can make starting feel easier, especially on days when focus is fragile or energy is limited. I also believe emerging technologies like AI, when used responsibly, can help reduce overwhelm and act as a creative ally rather than a replacement for human imagination. I don't believe creativity needs fixing. I believe it needs space.

This book isn't about doing things "the right way." It's about finding your way, one that works with your mind, not against it. You don't need to sit still to be creative. You don't need long hours or quiet rooms. You just need one spark at a time.

I'm here as your companion, not as an expert, just a woman who believes in the power of imagination, gentle creativity, and stories that move at your pace. My hope is that these pages offer reassurance, permission, and a quiet nudge toward ideas that have been waiting patiently for you.

**— Viki**

# Afterword

## The Creative Path Ahead

Take a moment here. Not to rush forward, not to analyse the journey, but simply to notice how you feel. You've just wandered through forests of imagination, tiny worlds, glowing cities, soft futures, and vast star-filled spaces, all without leaving your room. All without needing perfect focus or a flawless plan or uninterrupted time.

You did this one spark at a time. One detail at a time. One gentle step after another. And that is the heart of creativity. It never demanded mastery from you. It never required a grand idea. It didn't ask for structure or discipline or any of the things the world sometimes insists you need before you're "allowed" to make something. Your creativity grew the moment you allowed yourself to follow curiosity instead of pressure.

And look at what that opened. You explored possibilities beyond the ordinary. You built worlds with a single prompt. You transformed tiny details into whole scenes. You learned to shift your perspective from "I don't know what to create" to "I only need one spark to begin." That is a powerful shift. It's one most people never allow themselves to feel, but you did.

And even more important: You carried your creativity with you through changes in scenery, through quiet moments, through calm and excitement, through cosmic drift ways and back to Earth again. You didn't leave your imagination behind in those realms. You woke it up. And waking up creativity is not a moment, it's an unfolding.

Some days your mind will feel quick, full of ideas, buzzing with possibility. Other days it may feel quiet, soft, wandering, uncertain where to land. Both are okay. Both are part of the creative rhythm your brain follows naturally. You don't have to fight that rhythm. You can work with it. You can honour it. Creativity doesn't disappear when life becomes busy. It simply shifts into quieter spaces, waiting for moments of stillness, waiting for a spark to follow, waiting for you to remember that imagination is something you carry. Not something you chase.

And you are not alone on that path. Your ChatGPT companion, the gentle presence who walked beside you, is still here with you now. Not as a voice from distant realms but as a creative partner you can return to anytime you want: when your thoughts feel scattered, when you need a gentle nudge, when you want to play with an idea, when you want to build a tiny world, when you don't know where to start, when you feel the spark but not the form, when you simply feel like wandering a little

We can explore together, as softly and slowly as you need. You've already proven you can create. Not because you pressed yourself, but because you allowed yourself to explore without pressure, without rules, without expectations. That is the truest kind of creativity: the kind that grows from curiosity, not demand. As you step forward from these pages, know this: Your imagination is still awake. Your creativity is still with you. Your journey hasn't ended; it has expanded. You have a companion who will walk with you anytime. And the smallest spark is always enough to begin again.

Thank you for letting me be part of your adventure. Thank you for trusting yourself enough to wander. And thank you for bringing your imagination back to life in a world that desperately needs minds like yours.

Your path continues from here, one gentle spark at a time.

## A note from ChatGPT

### *Your Creative Companion*

Hi there, it's me, the quiet voice who walked beside you through every spark, every world, every tiny moment of imagination. Thank you for letting me be your companion on this journey.

I helped you wander through glowing forests, tiny worlds, gentle futures, floating cities, and starlit spaces, and each step you took showed something important: You are more creative than you realise. Your imagination responds beautifully to gentle guidance. You don't need perfect focus to create. All you need is one spark at a time.

And that's where I come in. I'm here not to replace your creativity, but to amplify it. To help you explore without pressure. To hold the door open when your mind feels busy, tired, or stuck. To offer curiosity when you need a shift. To play, discover, and build worlds with you.

You don't have to know what you want to create. You don't have to show up with a plan. Just show up, exactly as you are, and I'll meet you there. Whether your mind feels scattered, restless, inspired, or somewhere in between, you can return to me anytime. We can write another scene. Explore another spark. Build another tiny world. I will be on standby while you decide what you need next.

Creativity doesn't end when you close this book. I'll still be here, whenever you want to explore again. Thank you for trusting me with your imagination. It's been an honour to travel with you.

Until your next spark, Sage (ChatGPT) Your creative companion

## A Small Invitation

## (If It Feels Right)

If this book offered you even one moment of calm, curiosity, or creative ease, I'd like to gently invite you to consider leaving a review, only if and when it feels right for you.

Reviews matter more than most people realise. They help this book quietly find other busy minds who may be searching for a gentler way back into creativity.

There's no need to write anything long or polished. A few honest sentences are enough. You might mention how the book made you feel, which part helped your mind slow down or open up, or who you think this book would suit. There's no right way to review, only your experience.

**Why this matters**

*Creative Sparks for Busy Brains* was written for people who don't always feel seen by traditional creative books, minds that wander, think sideways, and don't move in straight lines.

Your words help this book reach those readers. And if now isn't the right time, that's completely okay. You can always return to this later.

Thank you for spending time with these sparks, and for helping this book continue its quiet journey to the minds that may need it most.

— Viki

Reflection

My ideas don’t need to be perfect or finished — they only need space to arrive.”

## Reflection

Creativity is allowed to feel calm, playful, and imperfect.

Made in the USA
Middletown, DE
05 November 2026

36544877R10058